Hampshire County Public Library
153 West Main Street
Romney, WV 26757

1/2019

PLANT POWER
POLLINATING PLANTS

by Karen Latchana Kenney

pogo

Ideas for Parents and Teachers

Pogo Books let children practice reading informational text while introducing them to nonfiction features such as headings, labels, sidebars, maps, and diagrams, as well as a table of contents, glossary, and index.

Carefully leveled text with a strong photo match offers early fluent readers the support they need to succeed.

Before Reading

• "Walk" through the book and point out the various nonfiction features. Ask the student what purpose each feature serves.

• Look at the glossary together. Read and discuss the words.

Read the Book

• Have the child read the book independently.

• Invite him or her to list questions that arise from reading.

After Reading

• Discuss the child's questions. Talk about how he or she might find answers to those questions.

• Prompt the child to think more. Ask: What did you know about pollination before reading this book? What more would you like to learn after reading it?

Pogo Books are published by Jump!
5357 Penn Avenue South
Minneapolis, MN 55419
www.jumplibrary.com

Library of Congress Cataloging-in-Publication Data

Names: Kenney, Karen Latchana, author.
Title: Pollinating plants / by Karen Latchana Kenney.
Description: Minneapolis, MN : Jump!, Inc., [2018]
Series: Plant power | Audience: Age 7–10.
"Pogo Books." | Includes index.
Identifiers: LCCN 2018005682 (print)
LCCN 2018008594 (ebook)
ISBN 9781624968884 (ebook)
ISBN 9781624968860 (hardcover : alk. paper)
ISBN 9781624968877 (paperback)
Subjects: LCSH: Pollination—Juvenile literature.
Classification: LCC QK926 (ebook)
LCC QK926 .K44 2018 (print) | DDC 571.8/642—dc23
LC record available at https://lccn.loc.gov/2018005682

Editor: Jenna Trnka
Book Designer: Molly Ballanger

Photo Credits: Tsekhmister/Shutterstock, cover; Nataliia Melnychuk/Shutterstock, 1; Stone36/Shutterstock, 3; William Harvey/Alamy, 4; Arto Hakola/Shutterstock, 5; paulbein/iStock, 6-7; cam3957/Shutterstock, 8-9; Anton-Burakov/Shutterstock, 10; Gerard Lacz/age fotostock/SuperStock, 11; amenic181/Shutterstock, 12-13; jps/Shutterstock, 14; Jonathan S. Blair/Getty, 15; Danita Delmont/Shutterstock, 16-17; Petr Simon/Shutterstock, 18-19; tinglee1631/iStock, 20-21; Sari ONeal /Shutterstock, 23.

Printed in the United States of America at Corporate Graphics in North Mankato, Minnesota.

TABLE OF CONTENTS

CHAPTER 1
Nectar and Pollen 4

CHAPTER 2
Growing Seeds 10

CHAPTER 3
Pollinators 14

ACTIVITIES & TOOLS
Try This! 22
Glossary 23
Index 24
To Learn More 24

CHAPTER 1

NECTAR AND POLLEN

A hungry beetle lands on a flower. It wants the **nectar** inside. Nectar is liquid sugar. It gives creatures **energy**.

The beetle takes something else, too. What? Tiny grains of **pollen**. They stick to the insect's body. The beetle doesn't know it. But it will carry this pollen with it. It will **pollinate** the next tasty plant it lands on.

pollen

stamen

What is pollination? It is a process with steps. While a visitor eats from a flower, it rubs against the **stamens**. These are the male parts of the flower. Their tops are covered with sticky, powdery pollen.

The stamens surround the **pistil**. This is the flower's female part. The **ovary** is at the bottom. It will hold the plant's seeds.

At the top is the **stigma**. It is very sticky. Pollen from the wind or a hungry visitor easily sticks to it. Once pollen is on the pistil, it changes. The pollen grows a tube. It goes all the way down the pistil. This is how the pollen's male **cells** travel to the ovary.

pistil · · · · ▶

TAKE A LOOK!

A plant has both male and female parts. Pollination brings cells from these parts together. Then the plant can make seeds in the ovary.

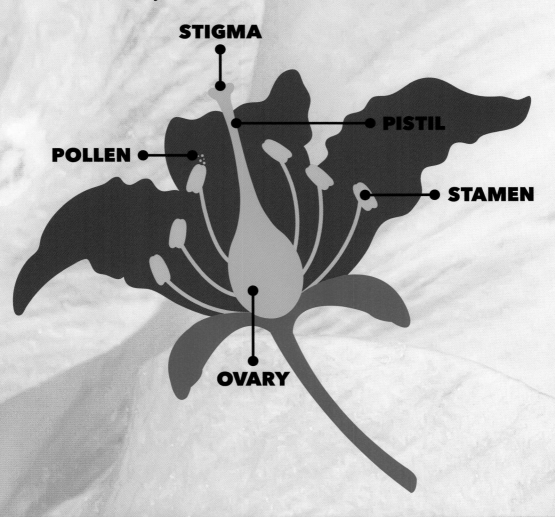

STIGMA

PISTIL

POLLEN

STAMEN

OVARY

CHAPTER 2

GROWING SEEDS

The male cells enter the egg.
The flower's petals drop off.
The ovary becomes a fruit.
It covers the growing seeds inside.

The fruit grows bigger and tastier. It attracts creatures. Soon animals will eat the fruit.

The animals carry the seeds in their bodies. They leave the seeds with their waste. The tiny seeds grow into new plants in new places. When **pollinators** find them, these plants grow seeds, too.

DID YOU KNOW?

Some seeds don't need to be eaten to travel. Some stick to animals or people. Some float in the water to new places. And some are carried by the wind.

CHAPTER 3
POLLINATORS

Insects and birds are big pollinators. But some unusual animals pollinate plants, too.

honey possom

The honey possum lives in Australia. This tiny possum eats nectar and pollen. It has a long snout and tongue to reach inside flowers. It passes pollen as it eats.

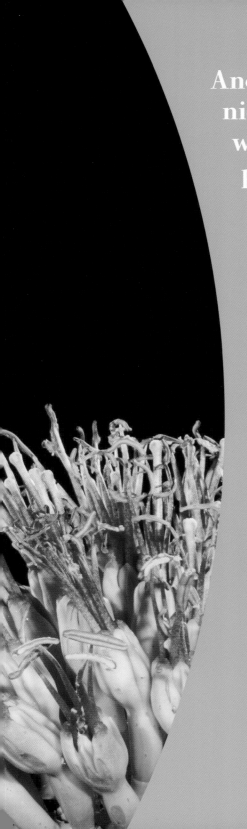

Another creature pollinates at night. This **mammal** flies with wings. It is a bat. What kinds of plants does it eat and pollinate? Bananas, mangoes, and many other fruit plants.

Some flowers have many visitors. But other flowers have nectar that is hard to reach.

Some hummingbirds have beaks made just for certain flowers. The sword-billed hummingbird has a very long beak. It feeds from long, tube-shaped flowers.

Bees are busy. They are the best insect pollinators. They fly from flower to flower. They gather nectar and pollen as food. They carry pollen as they do so.

DID YOU KNOW?

Bees are very important. Why? They pollinate a lot of the food we eat. How much? One out of every three bites!

ACTIVITIES & TOOLS

TRY THIS!

BUTTERFLY FEEDER

Make your own nectar to see how its attracts butterflies and other pollinators.

What You Need:
- brightly colored plate
- colored sponges
- scissors
- 1 cup water
- ¼ cup sugar
- pot
- spoon

❶ First make some nectar. Have an adult help you with this step. Put the water and sugar in the pot. Bring it to a boil. Then simmer and stir until you cannot see the sugar. Let the nectar cool.

❷ Cut the sponges into circle or flower shapes.

❸ Soak the sponge shapes in the nectar.

❹ Set the sponges on the plate. Put the plate outside.

❺ Check on the plate to see if any butterflies visit. How many do you see? What colors are they? Do any other pollinators stop to eat the nectar?

GLOSSARY

cells: The smallest units of a plant or other living thing.

energy: The ability to do work.

mammal: A warm-blooded animal with a spine and hair or fur; females make milk for their babies.

nectar: A sweet liquid in flowers that insects, birds, and other creatures drink.

ovary: The part of a flowering plant where seeds form and grow.

pistil: The female part of a flower.

pollen: Tiny grains on the male part of a flower that contain male cells to make seeds.

pollinate: To move pollen from one flower to another so that it can make fruit and seeds.

pollinators: Agents, such as insects and birds, that pollinate.

stamens: The male parts of a flower.

stigma: The sticky tip of the pistil.

INDEX

bats 7, 17

bees 21

beetle 4, 5

birds 7, 14, 18

cells 8, 9, 10

colors 7

egg 10

energy 4

flower 4, 7, 8, 10, 15, 18, 21

fruit 10, 11, 17

honey possum 15

insects 5, 7, 14, 21

nectar 4, 7, 15, 18, 21

ovary 8, 9, 10

pistil 8, 9

pollen 5, 7, 8, 9, 15, 21

pollinate 5, 14, 17, 21

scents 7

seeds 8, 9, 10, 13

stamens 7, 8, 9

stigma 8, 9

sword-billed hummingbird 18

water 13

wind 8, 13

TO LEARN MORE

Learning more is as easy as 1, 2, 3.

1) **Go to www.factsurfer.com**

2) **Enter "pollinatingplants" into the search box.**

3) **Click the "Surf" button to see a list of websites.**

With factsurfer, finding more information is just a click away.